OUR EARTH

P K N Panicker

MINERVA PRESS
NEW DELHI
MIAMI LONDON

OUR EARTH

ISBN 81 7662 135 8

First Published 2001 by
MINERVA PRESS
13 Palam Marg
Vasant Vihar
New Delhi – 110 057

Printed & bound in India
for Minerva Press (India) Pvt. Ltd.

OUR EARTH

Preface

An exposure, in a very general way about various aspects about our place of living – the earth, such as:

a. Certain geological and geophysical data.
b. The genesis and evolution of life.
c. Man's civilisation and its development through stages.
d. Environment degradation and man's responsibility, to growing children in their formative years, is considered necessary. The target group is children in the age group of 10 and 15 years and it is hoped that the contents will, to a certain extent, satisfy the inquisitive minds of school children who have many questions regarding their surroundings.

Many published works are available on the subjects covered. However, they specialise in one or other of the various topics listed above and most often are voluminous – each running into 400 or more printed pages. It is felt that children cannot spend time to study such books and equally they will not have the patience. Hence, a small book, covering all the topics and at the same time comprehensive, is considered appropriate. This one is planned for three-hours' reading.

While dealing with the developments during the 20th century, I have not gone into details; rather thought it is sufficient to just mention the important events and leave it to the students to look for more details elsewhere.

P K N Panicker

Contents

Chapter I
BEAUTIFUL EARTH

In the mornings when we get up from sleep we are greeted by newly blossomed flowers around us. The birds, ready to leave the nests in search of food for themselves and their young ones, greet us with their melodious music. The trees and plants around us greet us, waving their branches and leaves. And we, men and women, boys and girls greet each other with a 'Good morning'. Yes, this earth on which we live is undoubtedly a beautiful place!

When we go to the beach and watch the waves of the ocean, we wonder where the vast spread of water ends on the other side. When we stand by snow-covered mountains, we feel we can climb and climb till we touch and kiss the beautiful sky above us. Oh, how big and vast is our Earth!

But what is this earth really like? Is it as big as we tend to think? Is it as beautiful as we imagine? What is it made up of? What is the shape of this earth on which we walk and trample with our feet? Is it not natural that we are curious to know more about the place we live on, the place, of which we talk, love and dream?

The fact (which we presume) is that our earth was not always as beautiful as it is now. In the beginning, it was a ball of very hot, flaming gases. Astronomers assume that all planets around the sun including our earth were formed at the same time; when the sun was quite young, about four and a half billion years ago. Such conclusions have been arrived at by scientists with precise measurements of radioactivity, and we have every reason to believe that the conclusions are reasonably accurate.

Chapter II

THE SOLAR SYSTEM

Our solar system, as we commonly understand it now, has nine planets revolving round the sun, in definite paths known as orbits. We also know that there are other solar systems other than ours, each with their own suns and planets around them. However, let us not make the subject more complex and stick for the present to our solar system, and more particularly to our earth.

Chapter III

SOME DETAILS ABOUT THE PLANETS

Name of Planet	Average Distance from Sun million km	No. of Satellites	Symbol	Diameter	Period for One Rotation round Sun	
					Years	Days
Mercury	57.6	None	☿	4800	0	88
Venus	107.2	None	♀	12320	0	225.
Earth	148.8	1	⊗	12640	1	0
Mars	227.2	2	♂	6720	1	322
Jupiter	772.8	12	♃	142400	11	315
Saturn	1417.6	10	♄	12000	29	167
Uranus	2852.8	5	♅	48000	84	7
Neptune	4468.8	2	♆	44800	134	284
Pluto	5865.6	None	e	14400	248	120

The orbits or the paths through which the planets go round the sun are not exactly circular, but are slightly oval shaped and are called ellipses. The planets always move around the sun and cannot escape from its gravitational attraction. Our earth is just one planet among the nine in the solar system. We often think of our earth as the most important body in the universe. However, astronomers tell us that there is nothing very special about the earth, except that the human race lives on it. Some astronomers think that there are many other planets similar to the earth in the planetary systems of certain other stars other than the sun, but so far we have no definite means of knowing whether life exists on any of them.

Apart from the planets, there are billions and billions of smaller objects in space. Some of them, which are very much smaller than the smallest planet, mostly of diameters varying between 15 and 80 kilometres (the larger ones among them are – Pallas – 490 km; Vesta – 390 km; Juno – 190 km; and Ceres – 77 km) go around the sun in their specific orbits like other planets.

These are named asteroids. Others of importance in the solar system are the comets and meteors. Comets have very large sizes but they are of much lesser weight than the earth. They look like objects with long tails, which in fact are either vaporised or ionised material. Comets are generally very bright and go round the sun in flattened ellipsoidal, parabolic or hyperbolic orbits. Meteors are very small objects moving round the sun. Occasionally, they fall apart from their orbits and get into the earth's atmosphere. Very rarely an asteroid or a comet may enter the earth's atmosphere and fall on the earth's surface. There have been a few incidents in the past, and objects thus fallen have left huge craters on the earth's surface for us to see even today. However, a large number of small meteors fall into earth's atmosphere every day. It is estimated that around 20,000,000,000,000 ($2x10^{13}$) meteors, each weighing less than 0.0001 gram (total mass of 1000 tons) enter the earth's atmosphere every day.

Chapter IV
ORIGIN

The earth did not always exist. Scientists believe that it was created at the same time as other planets and the sun. There are many theories about the origin of the solar system. According to the most widely accepted theory, the sun and the planets were formed out of an immense cloud of fine dust-like material, comprising cosmic gas and dust spread over several million kilometres, which was pulled together by gravitational forces into large chunks of mass – which we now call the sun and the planets. The earth is one such mass.

The mass that is earth was compressed under its own weight and again grew hot. Radioactivity also contributed to the heating and it became almost like a semisolid molten mass. At the same time, the interior of the mass got compressed under enormous pressure.

Rotation gave it the present nearly spherical shape or, to be more precise, the spheroid shape. It is slightly flattened at the poles. The polar diameter is 43.2 km short of the equatorial diameter of 12681.6 km.

Chapter V
BIOSPHERE

The earth presently consists of an inner layer or core believed to be made up of heavy metals like iron and nickel, probably in a solid state. It is about 2720 km in diameter. The core next to the innermost one, which is about 2,080 km thick probably consists of the same metals as in the inner core but in a molten liquid state. Outside this core lies the middle layer, a solid mantle of rock, about 2860 km thick. The crust or the solid outer layer is about 40 km thick. Upon and within the crust lies the hydrosphere or the moist envelope, which includes surface waters and water-saturated soil and rock. The biosphere, or the home of all living things on earth, extends a short distance up into the atmosphere and down into the hydrosphere.

Chapter VI

BEGINNING OF LIFE

When the earth finally condensed into the solid spheroid that it is today, the atmosphere enclosing it probably contained varied amounts of hydrogen (H_2), water (H_2O), ammonia (NH_3), Methane (CH_4) etc. Under the influence of ultraviolet radiation from the sun many chemical changes took place. In the process, ammonia got converted into water and nitrogen. Iron and other metals on the crust got converted into their respective oxides and sulphides. Also, the continued chemical processes that were taking place on the earth and its atmosphere resulted in the synthesis of many organic compounds. Among them those which ultimately led to the evolution of life, were the complicated and big molecules of amino acids and simple proteins, which are rightly known as the building blocks of living substances.

Chapter VII

THE WATER CYCLE

The earth was originally a ball of rock and metals like iron and weighed around six thousand million million million tonnes (6×10^{21}), that is 6 followed by 21 zeros). The atmosphere surrounding it cooled over the years. As a result of cooling, the earth's crust shrank and shrivelled unevenly and formed mountains and valleys as well as large cracks and hollows. Simultaneously, when the atmosphere cooled sufficiently, the water vapour in the atmosphere condensed and resulted in rain, snow and hail. On the other hand, part of the water that accumulated in the cracks and hollows evaporated due to radiant heat from the sun and escaped into the atmosphere where it again cooled and condensed and fell back to the earth's surface as rain and snow. Thus, a constant circulation of water between the earth and the atmosphere started. In this way, rivers, lakes, seas and oceans were formed. Through the ages, the earth received a lot of rainfall, and now most of the earth's crust is covered with water. About 7/10th of the earth's surface is covered with water and 3/10th of it is land forming different continents and numerous islands.

Over 97 per cent of all the earth's water is salty. The remaining 3 per cent is fresh water, of which almost 75 per cent is in the form of polar ice caps and about 10 per cent to 15 per cent is deep below the earth's surface and hence is not usable. Only about 0.3 per cent of all the earth's water is in the soil, in wells, rivers and lakes – and all life on land depends on this. However, the constant circulation of water between the earth and its atmosphere keeps the balance of non-salty water on the land and thus depletion of fresh water is prevented. Therefore, the process of constant circulation of water is very essential to sustain life, which needs non-salty water.

Chapter VIII

INTERIOR OF THE EARTH

The landmass forming the various continents is made of silica and aluminium known by the abbreviated name sial. Ocean floors are made of silica and magnesium known as sima. In other words, sial covers the surface of the landmass, whereas sima is generally down below. For example, the roots of the Himalayas, with the peaks 8.8 km above sea level, will probably be between 50 and 60 km below. At the foot of the Himalayas, the great Indian Plains have only a thin layer, say about 2.0 km thick of sial. Below that is a great mass of sima, probably a layer of 50 km or thicker.

Chapter IX
VOLCANOES

The interior of the earth is hotter than its surface. When mines or deep wells are dug, it is found that the temperature down below is higher than at the surface. The increase in temperature as we go down varies between 10°C to 50°C per km, depending on the place. At a depth of 40 km, temperature is estimated to be around 400°C, at 200 km 1500°C and at 2900 km 3000°C. Naturally, at these depths and below, sima is expected to be in a molten condition. Molten sima, which is under great pressure due to the mass above, pours out like a jet, just as water under pressure pours out from a broken pipe, through weak spots or cracks on the surface of the earth. The molten sima, which pours out like this, is called lava and the vent through which this sudden pouring out occurs is called volcano. The number of known active volcanoes is 600, out of which 109 are under water in different oceans or seas.

Chapter X
EARTHQUAKES

The crust of the earth is continuous but is not uniformly thick. The crust can be assumed to be floating on the surface of molten sima. Numerous adjustments take place often in the lower layers beneath the crust, resulting in parts of the crust at times sinking and at other times rising. These adjustments result in vibrations or quakes on the crust. Sometimes, such adjustments result in fractures, or faults, as they are generally known, on the surface of the earth. Portions of earth sometimes slip along the fault and result in quakes. These may be considered the primary cause for earthquakes, though there may be other reasons including – man-made causes. Basins of large dams when filled, especially during the first few years after construction, can cause earthquakes. Explosions of large nuclear devices produce large shock waves and can be detected anywhere on the earth. Thus, we are in a position to even create artificial earthquakes and study these to gain a better understanding of the occurrence and behaviour of natural earthquakes. The science that studies earthquakes is known as seismology and the equipment used to study earthquakes is called a seismogram. Earthquakes are part of a continuing natural process, though sometimes they may be due to man-made causes. It would not be an exaggeration to say that the earth never rests.

Chapter XI

GLACIERS AND ICE CAPS

In Polar Regions and at high altitudes the summer is comparatively cool. In such places ice formed during the winter season does not completely melt in summer. The ice thus accumulated year after year forms large and very large ice blocks. Such big ice blocks formed at very high altitudes are called glaciers. The glaciers occasionally drift down because of their own weight and the speed at which they move down varies from a few centimetres to twenty meters per day, depending on the slope. Such large glaciers are found in the high altitudes of the Himalayas, which slide down the slopes, melt and cause floods in the Ganges. However, most of the world's ice is in Antarctica which is an area almost 2.5 times that of India, and is covered with ice to a depth of more than 2300 metres. Such large spreads of ice are known as ice caps. Similar ice caps are found in the Arctic region also; in Greenland, Iceland and many other places. Large quantities of water have been transformed into ice, resulting in a drop of the sea level by approximately 130 metres. If today all the ice were to melt all of a sudden, many of our well-known coastal cities and ports would go under the sea. However, the chances of this happening are negligible and we can be happy about it.

Chapter XII

RIVERS

We have already seen how the water cycle works. While a part of the large quantity of water that falls on the land mass as rain evaporates, the rest of it seeps into the soil and rocks and a part of it slips down hills and planes and reaches the oceans. When water slips down the hills and plains at different points and flows in the same direction, the chances are that some would meet and combine to form bigger streams and by the time such streams reach the plains they become mighty streams of water known as rivers. Most rivers flow in snake-like zigzag courses, depending on the topography. Also, the force with which water flows down the hills and plains is such that it erodes the soil and cuts courses on its own. Thus, we can definitely say that water plays an important role in moulding the shape of the land.

The longest rivers in the world are:

Name	Continent or Country	Length km
Nile	Africa	6632
Amazon	South America	6480
Mississippi-Missouri	North America (USA)	6016
Irtysh	Asia	5120
Yangtze	Asia (China)	4960
Amur	Asia	4640
Congo	Africa	4349
Hwang Ho (Yellow)	Asia (China)	4320
Lena	Asia	4232
Mackenzie	North America (Canada)	4216
Mekong	Asia	4160
Niger	Africa	4160

Name	Continent or Country	Length km
Yenisey	Asia	3776
Murry-Darling	Australia	3960
Volga	Europe	3928
Medira	South America	3366
Parana	South America	3286
Sindhu (Indus)	Asia	3168
Yukon	North America (USA)	3166
St. Lawrence	North America (Canada)	3112
Rio Grande	North America (USA)	3016

Chapter XIII

OCEANS

It has already been pointed out that 70 per cent of the earth's surface is covered with water. In fact, 60 per cent of the northern hemisphere and 80 per cent of the southern hemisphere are covered under vast stretches of water known as oceans. The total area covered by the oceans is estimated to be around 357,000,000 sqkm.

In relation to oceans, we also use terms such as sea, for example the Arabian Sea; bay – example - the Bay of Bengal; gulf – example - the Persian Gulf; and lake – example - Lake Victoria. What do these different terms really mean? An ocean is a vast expanse of salt water surrounding the land areas, generally containing about 3.5 per cent of dissolved sodium chloride and very small quantities of magnesium, calcium and traces of other salts and minerals. The average depth of the ocean is about 4000 km and in certain places the depth is about 11000 km. The density of salt water in the ocean is normally 1.026, whereas that of pure water is 1.0. For our convenience, we have divided the ocean into five separate oceans and have named them as The Atlantic, Arctic, Antarctic, Pacific and Indian Ocean. A sea is also a great body of salt water but is smaller than the oceans and is generally landlocked. A gulf is a part of an ocean or sea that extends into the land. We can also say that a gulf is a landlocked sea larger than a bay. A bay is thus understood as a landlocked sea much smaller in size, compared to a gulf. A lake is an expanding part of a river or a reservoir formed by a dam. A lake generally has water without salt. However, the words sea, gulf and bay are used by practice and are not always according to the definitions given. For example, The Baltic Sea is smaller than The Hudson Bay. Likewise, some of the lakes, like Eyre in Australia and Rudolf in Kenya have salt water.

Details about the oceans and some seas

Name	Area sqkm	Avg. Depth km	Greatest Depth km
Pacific	1,63,804,160	4254	10969
Atlantic	80,716,800	3903	9134
Indian	72,576,000	3940	6960
Arctic	14,186,496	1273	5712
Antarctic	3,280,000	-	-
Mediterranean Sea	2,931,200	1364	4374
South China Sea	2,291,200	1636	4987
Caspian Sea (Lake)	4,35,200	Salt water, bordering Iran	
Superior (Lake)	81,460	Fresh water, bordering USA & Canada	

According to one classification, all seas and lands around the South Pole which lie within the Antarctic Circle at 66°33' are called the Antarctic Region and the sea between the Antarctic Circle and a latitude of 60° is called the Southern Ocean.

Large areas of rock surfaces within the Antarctic lie beneath ice caps, several thousand meters thick. Some of these rock surfaces are actually below sea level. The total area of land and sea within the Antarctic Circle is 11,520,000 sqkm. Approximately 10 per cent of the surface area of the earth lie under ice caps.

Chapter XIV
MOUNTAINS

As explained earlier, the earth's crust is floating on liquid sima and the thickness of the crust is not uniform at all places. Therefore, the mass or weight of the crust at different places is different. Imagine a shallow trough with water covered with a flexible plastic sheet. If more pressure or weight is applied at one point on the sheet, the water will give way, allowing the sheet at that point to go down, but the water thus pushed down will in turn push the sheet at some other point and an upward bulge will result. It is possible that at some point along the crust such bulges have occurred and these could be the origin of some mountain ranges. The Urals in Europe and the Appalachians in North America are examples of such mountains and were formed over a 100 million years ago. Likewise, some continents or landmasses, which are actually floating, could have drifted. These could have collided or pressed each other, resulting in folds, which naturally appear as mountains. It is believed that some of the most impressive mountain ranges of today, such as the Himalayas and the Alps are results of such continental drift. Since the process of drift is continuing, the height of the folds is increasing. Actual measurements have shown that the Himalayas are still growing. These mountains are younger and are about 40 million years only. Some details on important mountain peaks are given.

Details about some Mountain Peaks

Mountain Peak	Ht. Metres.	Location
Everest	8847.73	Nepal-Tibet
Kanchenjunga	8585.60	Nepal-Tibet
Dhaulagiri	8171.69	Nepal
Annapurna	8078.42	Nepal
Nanda Devi	7816.60	India
Namcha Barwa	7728.20	Tibey
K2	8610.60	India-China
Makalu	8474.56	Nepal-Tibet
Nanga Parbat	8125.97	India
Gosainath	8013.50	China
Rakaposhi	7787.64	India
Gurla Mandha	7723.20	Tibet

The science dealing with mountains is called orology or orography.

Chapter XV

SOME PHYSICAL DATA REGARDING EARTH

1.	Estimated age	4,500	million years
2.	Equatorial Diameter	12681.6	km
3.	Surface area	504,192,000	sqkm
4.	Land surface	147,225,000	sqkm (29per cent of total)
5.	Water surface	356,966,000	sqkm (71per cent of total)
6.	Circumference along the equator	39,843	km
7.	Circumference along the meridian	39,776	km
8.	Volume	1.064×10^{12}	cubic km
9.	Mass or weight	6.586×10^{21}	Tons
10.	Time for one revolution around the sun	1 year	365 days 5 hrs 48 mts and 46 seconds.
11.	Speed of revolution	1,06,560	km per hour
12.	Time for the rotation on its axis		23 hrs and 56 minutes.
13.	Speed of rotation	1,600	km per hour

The sun is the source of all light and heat in our solar system. The sun's energy is generated by nuclear reactions in its interior, which raises the temperature of the sun's core to about 140,00,000°C. The entire mass of the sun is probably in a gaseous state. Even though the sun is losing energy continuously (equivalent to a weight loss of four million tons per second), it is estimated that the sun will survive for at least another 8000 million years. Of this, about 63,000 tons of mass falls on the earth every year in the form of energy. Light from the sun takes about eight minutes to reach the earth.

The earth has one satellite, the moon. The moon has a diameter approximately one-fourth that of the earth – to be more precise say 3456 km and its surface area is approximately equal to the area of the Atlantic Ocean. The moon continuously rotates round the earth and it takes approximately 27.3 days to complete one revolution.

As there is a difference in the speed at which the earth rotates around its own axis and the speed with which the moon rotates around the earth, we are not able to see the moon in full on all days. The time between one new moon and the next is about 29.5 days. Since the moon does not rotate around its own axis, we are not able to see the other half of the moon. However, man has succeeded in taking photographs of the other half of the moon, which we are not able to see, from the artificial satellites put in space by man. The average specific gravity of the earth's mass is 5.6, that of the moon 3.3 and that of the sun 1.4. Because of the lower density and mass, the gravitational pull on the moon is only about one-sixth that on earth. Consequently, the weight of any article on the moon will be only about one sixth of its weight on earth.

Chapter XVI
CONTINENTS

The shapes of various continents (the land masses which are not submerged under the oceans and seas) were not the same as we see them today. In all probability, as suggested by the coastal configurations of various continental blocks, there was a time when most of the landmasses, which were not submerged under the oceans, formed one single block. Gradual changes occurred and perhaps around 450 million years ago, North America, Greenland, Scandinavia and the Russo-European area were all together as a North Atlantic continent; Africa and India were together as a large piece of land mass, and Australia and some other areas were joined together as an Austral-Equatorial continent. All these land masses were connected through stretches of exposed land corridors – the North Atlantic continent through partly submerged South America with Africa, and so on. It is believed that around 300 million years ago the shape of the continents underwent substantial changes. The North Atlantic continent remained nearly the same, whereas a new formation connecting South America, Africa, Arabian Peninsula, India, Indonesia and Australia emerged as a Southern continent or Gondwanaland. It was only some period between 7 million and 2 million years ago that the continents got more or less the same formation as we see today. The landmasses, which are not submerged, are grouped into continents and islands, mainly based on the area. The smallest continent is Australia. Landmasses of smaller area are classified as islands. There are seven continents and their areas are given below.

Name	Area in sqkm
Africa	18,708,800
Antarctica	8,24,0,000
Asia	27,108,800
Australia	4,748,654
Europe	6,485,280
North America	14,984,000
South America	11,008,000

Chapter XVII

AGE AND DEVELOPMENT

The age of the earth is something difficult for us to know correctly. However, as in all other things, man's curiosity has made him look at various methods to assess the age of the earth, if not correctly, to some degree of accuracy. Now, it is generally accepted that the age of our earth is in the region of 4,500 to 4,550 million years. Of this, the solidification of the earth's crust took nearly 1,000 to 1,500 million years and nothing much is known about that period. However, geologists have traced the history of the earth during the next 3,000 million years, mainly based on fossilised relics of plants and animals that many of the rocks contain. The earliest fossils found are those of bacteria-like organisms, and date back to around 3,000 million years. The course of evolution was not smooth or unbroken. Some plants and animals might have evolved only to die out millions of years later, whereas some others persisted to live unchanged; and some others continuously underwent changes leading to higher forms of life. Fossils of primitive marine plants and animals including seaweed and invertebrate creatures could be traced to as far back as 600 million years. Primitive fish-type animals seem to have evolved a 100 million years later. Land plants established themselves about 400 million years ago and amphibious animals around 350 million years ago. Mammals, of which man is the most advanced, date back to 250 million years, and modern man around 10,000 years.

Disturbances of the earth's crust, mainly due to the

continual movements and collisions of its internal layered segments, caused important changes in geography, climate and even the shape of the land and in turn influenced the evolution and distribution of animal and plant life. Certain disturbances that started around 26 million years ago and lasted for over 19 million years were responsible for the formation of the Alps and the Himalayas. Similar disturbances that occurred around seven million years ago and lasted for five million years resulted in the formation of the North Sea, the Black Sea and the Caspian Sea. While geologists have traced the age of the earth to almost 4,550 million years, historians, relying on archaeological discoveries and written records of ancient civilisations, have traced back 6,000 years of history – and this history is the history of the modern man as we know it today.

Chapter XVIII
THE TREASURES

The earth has many treasures stored in its crust as well as within it. The treasures on its crust are mostly in the form of rocks. Rocks are made of minerals, which consist of natural elements. A few elements such as gold and platinum are found in a pure state. But a majority of the elements are found in chemical combination with one or more other elements. Oxides are produced when metals combine with oxygen and sulphides when metals combine with sulphur. Minerals are formed in various ways, for example, by crystallisation from molten lava, by crystallisation from vapours, by cooling of vapours around active volcanoes etc. Over 2,000 minerals have been identified. It is unlikely that we may find newer minerals on the crust as we have done extensive exploration work. However, the possibility exists that as we go deeper we may find new minerals that are stable at the high pressures and temperatures, nearer the centre of the earth. Equally, when we are able to explore the moon and other planets more thoroughly, many more minerals, which are not known on earth, may be located. A list of some of the well-known minerals and their principal chemical components is given. Apart from what is shown in the list, there are many more minerals of great commercial value. Petroleum and natural gas are two such. We use both of these extensively as fuel for generation of electric power; as fuel to generate the necessary heat (often in the form of steam) to run our industrial establishments; as fuel to run our trains, commercial vehicles and cars; and also as feed stock (raw material) for the manufacture of a variety of chemicals and chemical products such as fertilisers, pesticides, weedicides – used in agriculture; synthetic fibres like terylene, nylon, polyester – used for clothing; plastics like polythene, poly vinyl chloride (PVC) – used in the manufacture of a large number of articles for day-to-day use etc., and also many drugs essential for the treatment of diseases. The statement, that today man's civilisation is greatly dependent on petroleum, is not an exaggeration.

Mineral & Principal Chemical Component

Galena	Lead Sulphide
Sphalerite	Zinc Sulphide
Cassiterite	Tin Oxide
Flouorite	Calcium Fluoride
Corundum	Aluminium Oxide
Azurite	Copper Carbonate
Chalcopyrite	Copper-Iron Sulphide
Malachite	Copper Carbonate
Pitchblende	Uranium Oxide
Olivine	Magnesium-Iron Silicate
Bauxite	Aluminium Hydroxide
Magnetite	Iron Oxide
Hematite	Iron Oxide
Torbernite	Copper-Uranium Phosphate
Cinnabar	Mercury Sulphide
Beryl	Beryllium-Aluminium Silicate
Garnet	Iron-Aluminium Silicate
Halite	Sodium Chloride
Common Salt	Sodium Chloride
Opal	Silicon Oxide
Quartz	Silicon Oxide
Brimstone	Sulphur
Coal	Carbon

Diamond	Carbon
Graphite	Carbon
Zircon	Zirconium Silicate
Kaolin	Aluminium Silicate
Apatite	Calcium Phosphate
Pentaldite	Nickel-Iron Sulphide
Ilmenite	Titanium Dioxide
Monazite	Thorium Phosphate

The abundance of various elements that make up the top layer of the crust known as the lithosphere, consisting the top 16 km of the surface of the earth, have also been estimated. The most important of them are listed. The density of the elements differs, and therefore the volume and weight are different. For example, oxygen is about 30 times the volume of silicon and 10 times that of magnesium of the same weight.

Element	per cent by wt.	per cent by vol.
Oxygen	46.60	91.97
Silicon	27.72	0.80
Aluminium	8.13	0.77
Iron	5.00	0.68
Calcium	3.63	1.48
Sodium	2.83	1.60
Potassium	2.59	2.14
Magnesium	2.09	0.56

Chapter XIX

THE ATMOSPHERE

The blanket of gases surrounding a planet is known as its atmosphere. On earth, it is the atmosphere surrounding it that makes life possible. We have already seen that the biosphere, or the home of all living things on earth, extends into the hydrosphere, that is the moist stratum of the crust and equally a short distance up into the atmosphere. In the absence of the atmosphere, harmful radiation from the sun would have made it impossible for any life on earth. Even the temperature would have been very high but for the atmosphere. Though it is very difficult to definitely state the height to which the atmosphere around earth extends, we can safely assume that it exceeds 1,500 km. In fact, the air just thins out and gradually the density (that is the weight of air in grams per cubic meter of volume) becomes the same as that in the space between planets or what is generally known as outer space. Most of the mass of the atmosphere is contained in the first 8 km from the earth's surface. We call this mass air, and the composition of air is given. The figures given are to be taken as average and minor variations do occur from place to place.

Composition of Air

Name	Symbol	per cent volume
Nitrogen	N_2	78.084
Oxygen	O_2	20.946
Argon	Ar	0.934
Carbon dioxide	CO_2	0.033

Name	Symbol	per cent volume
Neon	Ne	0.001818
Helium	He	0.000524
Krypton	Kr	0.000114
Xenon	Xe	0.0000087
Hydrogen	H_2	0.00005
Methane	CH_4	0.0002
Nitrous oxide	N_2O	0.00005

The composition given is of air free of water vapour. Water vapour is always present in air but the quantity varies from place to place and also according to seasons and weather conditions. Over desert regions the quantity of water vapour is very small but near the oceans it is high. Carbon dioxide exerts great influence on the climate because it absorbs large quantities of infrared rays coming from the sun. Carbon dioxide is also very important because plants need it to manufacture their food. Plants, under the influence of light, convert carbon dioxide and water into sugar, starch and oxygen and this process is known as photosynthesis. This reaction is also referred to as fixation of carbon. Carbon dioxide can thus be thought of as one of the prime components that helped the evolution of all living organisms on earth from the beginning. However, it should be clearly understood that excessive presence of carbon dioxide in the atmosphere is harmful to all forms of life. The balance of carbon dioxide in the atmosphere is maintained by plants. Similarly, it is only plants, which can produce primary food, which in turn is consumed by all animals including man. These are the reasons why plant life is important to us and the reason why we should learn to protect and love plants. For convenience, scientists have divided the atmosphere into different stratum. The stratum nearest to the earth's surface and extending up to 10 km is known as the troposphere. This is the stratum in which man lives, in which clouds form and where all changes in weather, as man sees it, occur. Air pressure at sea level is maximum at 760 mm of mercury, (1.0133 bars or 14.7 lbs per since.) and decreases with altitude. At a height of 10 km, the pressure will be around 250 mm of mercury only. Likewise, in this stratum temperature decreases with altitude. The temperature at sea level varies from place to place, but in tropical regions it is around 30°C and in temperate regions around 15°C; whereas at an altitude of 10 km the temperature will be around 60°C only.

The next stratum is the stratosphere, which covers a height of approximately 10 to 35 km. Throughout this stratum, the temperature remains almost constant at around 60°C. At the lower boundary of this stratum, weather conditions are as on earth's surface but rain-bearing clouds are absent. This is one reason why modern Jet aircraft fly above the troposphere and at the lower regions of stratosphere, at altitudes near about 12,000 metres.

The third stratum is the mesosphere, covering a height of 35 to 75 km. However, at an altitude of 30 to 50 km, a zone of high chemical activity exists. At the lower region of this zone, a thin layer of sulphate molecules is found. Immediately above this is a zone characterised by the accumulation of ozone gas, which is formed as a result of the action of ultraviolet rays from the sun on the molecular oxygen of the atmosphere. Thus, the ozone layer can be considered as protecting the earth from the ill effects of ultraviolet radiation by absorbing it. In this region the temperature is high, compared to other regions surrounding it. The temperature in this region is around 25°C. In fact, in the higher regions, say around 30 km from the surface of the earth, the temperature increases and at about 40 km. It again decreases touching a low of about 73°C at around 70km; and then again increases with altitude. The temperature at an altitude of 320 km could be as high as 1,000°C.

Above the mesosphere is the ionosphere, extending from 75 to 550 km. In this stratum the short wave radiation from the sun has a dominant effect. This stratum is characterised by the presence of a large number of electrically charged ions and free electrons. Here, temperature increases with height. The ionosphere is further designated as the D, E, F_1 and F_2 layers. The D layer, above about 80 km is present only during daytime, and is where sunlight breaks up water vapour into hydrogen and hydroxyl molecules. The E layer, between the D and F layers, is of particular importance as this reflects back long radio waves to the earth by the electrically charged particles. F_1 starts above 160 km and the F_2 layer above 250 km. Both these layers, together known as the F layer or the Appleton layer, reflect back short radio waves. The D layer is at the lowest level of the ionosphere.

The exosphere extends beyond 550 km and merges with outer space. Protons, alpha particles and the nuclei of heavy atoms enter the atmosphere in the form of cosmic rays from outer space and collide with atoms and molecules in the upper atmosphere. These collisions produce secondary cosmic rays, which infiltrate the lower reaches of the atmosphere. In the exosphere, particles move freely and some particles may even escape into outer space. We have still to learn a lot more about the atmosphere and particularly about the upper stratum. Man is continuously trying to understand more about the atmosphere, outer space and other planets and unravel the secrets thereof. That is why we are sending artificial satellites to study atmospheric conditions, and specially designed space ships to go near and around the planets. Such satellites are equipped to send us valuable information and photographs from space. Some satellites also help us to improve our understanding of the earth by sending back photographs of the earth taken from above. They also help improve our communication, entertainment and information systems, such as the television, by reflecting back radio waves to us from different corners of the world.

Chapter XX

CLIMATE

We normally understand climate as the average course or condition of weather at a given place over many years as exhibited by the average and extremes of temperature, wind velocity, rainfall, humidity and other weather elements. The climate of a place is mainly dependent on its latitudinal position on earth. In fact, climate is a complex phenomenon in which the quantity of heat the earth receives from the sun plays a very important role. The solar radiation that reaches the earth's surface is affected by the angle at which the sun's rays strike the ground and the length of the day. A place on or near the equator receives maximum heat from the sun, and less when the place is near the Polar Regions. This is because the slanted rays from the sun are scattered over more of earth's surface and also because they have to pass a longer distance through the atmosphere. This is the prime factor in determining the climate of a place. However, other factors like movement of air masses, the quantity of water vapour in the air (in other words, humidity) and geographical conditions such as nearness to seas, position with respect to mountains, altitude etc. also modify the climate of a place. Winds also play a major role, but winds themselves are largely caused by the thermal action of the sun. Irregular distribution of land and water masses also contributes to climatic differences at places along the same latitude. Solar radiation is absorbed by land but mostly reflected by water. Therefore, the oceans get heated up slowly; but hold the heat longer, compared to land. This also is a factor that modifies the climatic condition of a place. It is interesting to note that the North Pole is not the coldest place in the northern hemisphere. This is because of the location of the North Pole in the Arctic Ocean and the influence of the floating ice caps around the place. On the other hand the South Pole, which is at the centre of a continental land mass and at very high altitude, records temperatures very near to the coldest ever recorded (in Vostok, very near the South Pole in August 1960, the recorded temperature was – 197°C).

Based on broad climatic grouping, the earth is divided into zones. In this respect, the two hemispheres are almost identical, even though variations do occur. On either side of the equator and within the latitude of 23.5°, where the sun reaches its greatest declination north or south (along the line of the tropic of Cancer in the northern hemisphere and tropic of Capricorn in the southern hemisphere), lie the tropical zones. In these regions, the climate remains reasonably stable throughout the year and the variation in temperature is not appreciable. However, these regions normally have high temperatures. Areas, lying between the tropic of Cancer and the Arctic Circle in the northern hemisphere and the tropic of Capricorn and the Antarctic Circle in the southern hemisphere, are known as temperate zones. These regions experience appreciable seasonal changes in temperature but do not reach extremes. The two Polar Regions, where temperature drops to extreme levels, lie beyond the Arctic and Antarctic Circles.

Chapter XXI

LIFE IN THE SEA

The best way to understand how existing species of plants and animals evolved is to follow the path of palaeontology, a young science based on the estimation of the ages of fossils. Fossils are the remains of the physical features of life from the past, mostly preserved in association with rock stratum. Interestingly, by a study of the fossils, scientists reconstructed a fish species known as Undina, which was supposed to have existed during the Cretaceous period, that is between 65 and 136 million years ago. Later a fish, a living genus, coelacanth, was discovered and surprisingly the fish was very similar to Undina. This confirmed the accuracy of the reconstruction techniques.

However, in spite of all our studies and the knowledge gained so far, even today we are not able to know exactly when and how life originated on earth. Nevertheless, sufficient evidence exists to assume that some forms of life were present three billion years ago. Fossils of organisms, resembling single cell algae, have been found in rocks estimated to be that old by radioactive methods. It is certain that life originated in the sea and underwent a process of continuous evolution from then on.

The first step in the evolutionary process was the increase in the size of the cell and the second step, which was indeed a leap forward, was the occurrence of photosynthesis. This step, which would well be called the first green revolution, took place around two billion years ago and the green algae containing the remarkable substance chlorophyll appeared on earth. This enabled the organism to capture energy from the sun's light and to use this energy to perform a complex chemical reaction, making the production of carbohydrates, drawing out carbon from carbon dioxide and hydrogen from water. In the process, free oxygen was released. Thus, with this step, the foundation was laid for an ecological balance to maintain the oxygen level in the atmosphere and for the preparation of food that could support other complex life forms eventually leading to man. The next major step, perhaps, was necessitated as a consequence of the increase in the size of the cells to the evolution of cells with multiple nuclei; and later, to the division or mutation of cells to form multi-cellular organisms and establish a definite methodology of reproduction.

During the pre-Cambrian period, about 600 million years ago, vegetation in the form of seaweed was supposed to have originated. During the Cambrian period, probably 578 million years ago and lasting for about 70 million years, many invertebrates such as worms, Jellyfish, sponges and trilobites evolved. However, seaweed was the only plant that provided food for these animals. Between 500 and 435 million years ago, known as the Ordovician period, vertebrates seem to have evolved as evidenced by the fossils found in western USA. The Silurian period between 435 and 395 million years probably saw the beginning of life on land in the form of leafless plants. Coral reefs developed on a large scale during this period. At the same time, vertebrates like sea scorpions three meters in length possibly existed during this period. The Devonian period (395 to 345 million years), also known as the age of fish, saw the development of most ancestors of all modern fish. Primitive sharks, measuring up to six meters in length, were found during this period. Marine life, both plant and animal, abounded during the next stage, known as the

Carboniferous period (345 to 280 million years). However, the next Permian period (280 to 225 million years) marked the end of the dominance by marine creatures as animal and plant life on land increased. These six geological divisions, namely the Cambrian, Ordovician, Silurian, Devonian, Carboniferous and Permian periods, which lasted for about 345 million years are together known as the Palaeozoic Era and were characterised by an evolutionary explosion of plant and animal life.

The Triassic and Jurassic periods, each lasting about 32 and 57 million years respectively (225 to 136 million years), saw fish-shaped reptiles, flying fish and lobster-like creatures. Aquatic reptiles were rapid swimmers and preyed on fish and other marine creatures. The Cretaceous period (136 to 65 million years) saw giant turtles, flying reptiles and varieties of shark, closely related to the sharks of today. However, the next 27 million years or so represented by the Eocene and Palaeocene periods, saw the extinction of most marine reptiles; but two groups of mammals, the early whales and sea cows began to evolve themselves and adapt to sea life. This was the period in which most species of fish in the ocean took on the shape and form we know today. New species of crabs and snails evolved during the Oligocene period (38 to 26 million years). The Miocene period (26 to 7 million years) saw the evolution of giant sharks, measuring almost 18 meters in length, but the next period, Pliocene (7 to 2 million years) saw the extinction of these giant sharks. Marine life, both plant and animal, evolved during this period to forms and habits as we see them today. Only very marginal developments or changes seem to have taken place over the last two million years. The sea, where life began, contains almost every main category of animal life including mammals, the warm-blooded groups to which man belongs. There are about 15,000 known species of sea fish and 60,000 species of mollusc which include oysters, mussels etc. The food cycle in the sea depends on microscopic plants known as phytoplankton, the primary source of food for minute animals called zooplankton. In turn, zooplankton are preyed upon by larger animals which themselves are preyed upon by still larger ones. Thus, a ruthless and never-ending cycle of the bigger devouring the smaller ones goes on and on. As most plants need sunlight for their survival, most marine plants grow within 100 meters of the water surface and consequently a large part of sea life is confined to this top most layer.

Chapter XXII

LIFE ON LAND

As pointed out in the earlier section, it was during the Silurian period that the first vegetation in the form of leafless plants appeared on land. During the Devonian period, plants as tall as 15 meters in height were probably in existence. Different kinds of fern came into being and towards the end of this period the earth began to look green. With abundant vegetation to feed on, invertebrate animals started leaving the sea and adapted themselves to life on land. These include millipedes, mites, spiders and wingless insects. The tropical climate was admirably suited for the growth of evergreen forests. The Carboniferous period saw giant trees, reaching heights of 30 meters. Most creatures, which migrated to land, found many conditions that necessitated substantial changes in their body system and structure. Particular mention is necessary of the oxygen intake system as oxygen became necessary for the assimilation of food. Most animals in the sea had systems to absorb the dissolved oxygen in the water either through moist skin, special tissues or mechanisms like the gill. The development of the lungs, which enabled absorption of gaseous oxygen from air, was a major step in the evolution of animals on land. The next important aspects in this process of evolution were the support and locomotion of the animal's own body. In water, living aquatic creatures whose specific gravity is almost equal to that of water, as we know from Archimedes' principle, have no difficulty in floating. This is not the case with animals on land.

Another major step was the evolution of the reproductive process to suit conditions on land. In water, the liquid medium conserves the eggs and receives the young. On land, air does not offer these facilities. Therefore, amphibian adults lived on land but laid their eggs in water; and their young ones spent the first part of their lives in water. This process of adaptation continued till finally the higher forms of life such as the mammals evolved a reproduction system wherein conception and gestation occurred within the body of the mother. Here too, the development of the embryo took place in a liquid medium, which reminds us of the aquatic ancestral origin.

During the Carboniferous period, the reptiles started breeding on land. Also, this was the period when winged beings appeared in the form of winged insects. The predominance of marine life started declining as life on land started increasing during the Permian period. Animal life on the whole remained reptilian. Mammals – warm-blooded creatures – evolved during the Triassic period. Dinosaurs, just about 15 centimetres long, also appeared during this period. During the Jurassic period also, mammals remained small sized. However, reptiles increased in size and variety. For the first time, creatures which could be termed as birds evolved feathers and wings. Archaeopteryx – the first bird – retained many of the reptilian characteristics. Some reptiles, too large for survival on land (25 meters long and weighing 35 tons) took to swamps and marshes. The Cretaceous period saw giant reptiles, dinosaurs and pterosaurs dominating land, but sadly enough, the same period saw the extinction of dinosaurs and many other giant animals. Birds evolved into two distinct types; one with well-developed wings almost similar to modern birds; the other wingless but with strong developed legs for swimming like the sea birds. This was also the period when placental mammals, whose young are nourished directly by the mother's blood until birth, appeared. Ancestors of the present-day elephant, rhinoceros, horse, pig and cattle originated during the Eocene and Palaeocene periods. Probably, primitive monkeys and gibbons appeared in Burma (Myanmar). The Oligocene period was, perhaps, the one in which tail-less primitive apes, possibly related to the ancestors of man, appeared. The ancestors of modern cats, dogs and bears also evolved during this period. The next three periods were the ones, which saw the development of modern man. The Miocene period was noted for the development of the primitive anthropoid ape, (Proconsul) in Central Africa, and gibbon-like apes (Pliopithecus) in Southern Europe. The Pliocene period saw the ascendancy of man-like apes, which continued to develop and thrive. The species of apes known as Australopithecus, which walked upright in open countries, also appeared during this period. During the Pleistocene period (2 million years to 10,000 years) apes developed enough intelligence to make implements to cut the animals they killed. Thus, intelligent activities and a planned work culture could be counted as having started during this period. From then on, it was a continuous process of development of the brain and intelligent activities in contrast to the evolution and adaptation of physical forms and organs. The Holocene period which started about 10,000 years ago, saw the development of man as a social being and his struggle not to adapt himself to the natural conditions around him but to make efforts to utilise the resources of nature around him for his well-being and his attempts to bring the natural forces under his control. Man learnt to domesticate other animals and make them work for his benefit; he started cultivation of plants to yield fruits for his food. We can say that from then on to this day it was the story of man and his civilisation.

Chapter XXIII
EVOLUTION OF MAN

Over 350 million years ago, the invasion of land by the amphibians was the beginning of a very spectacular and decisive advance in the process of evolution and adaptation on earth. Life on earth began to change and diversify at a faster rate than before. Geographical changes were taking place almost simultaneously. Landmasses moved apart and almost took the shape we see today; the Alps and the Himalayas rose from the sea floor.

The advance of amphibians to reptiles, which freed them from dependence on an aquatic environment and further into two classes, birds and mammals, may be considered as the first step in the process of evolution towards modern man. Mammals have three general characteristics.

1. They are warm-blooded. In other words, their body temperature is independent of the environment temperature. This quality allows them to adapt to areas with divergent climatic conditions, with comparative ease.
2. They have a higher degree of sensitivity, which allows them to easily spot the movement of other animals.
3. Mammals are viviparous, that is, they give birth to perfectly developed young and after birth protect and nurse the young as long as needed to make them self-supporting adults.

These characteristics enabled them to adapt to situations. As a result, some mammals learnt to hunt in the air, like bats;

others, to live on trees, like the monkeys; some, to burrow under the ground, like the rodents. Still, others went back to the sea, like the whales. There are nearly 5,000 species of mammals in many varieties of shape and form. Among these are the primates, headed by man.

The development of tall limbs for general agility and movement, and that of big toes and opposable thumbs were characteristics of special importance associated with the primates. The sense of hearing slowly decreased and the power of sight increased. A tendency towards an upright posture was also seen in the primates. During the Palaeocene and Eocene periods, mammals also started developing intelligence. As intelligence showed signs of development, the capacity of the skull and the weight of the brain also increased. Another important development was the appearance of the hand, which may have occurred towards the latter part of the Oligocene period. Towards the end of this period, the differentiation between apes and the ancestors of man is considered to have started. Australopithecine, hominoids of small stature and brain, teeth and posture showing many human features, appeared around four million years ago and were the first anthropoids to be classified as man.

An approximate idea of the evolution of man (a highly advanced primate) from the stage of the earlier, less advanced species is given in a tabular form.

	Period	**Stages of Evolution Leading to Modern Man**	**Found In**
Miocene	26 m yrs. ago to 7 m yrs.	Differentiation of apes from ancestors of man began. Early primates, lemurs, monkeys, apes and gibbons.	East Africa India etc.
Pliocene	7 m yrs. ago to 4.5 m yrs.	Oreopithecus (ape) Pliopithecus (Gibbon)	Asia Italy

	Period	Stages of Evolution Leading to Modern Man	Found In
Lower Pleistocene	4.5 m yrs. ago to 500,000 yrs.	Homo habilis Australopithecine Olduvai Pithecanthropus (Java man)	East Africa South Africa East Africa Java
Middle Pleistocene	500,000 yrs. to 100,000 yrs.	Veretesszollos man Heidelberg man Pithecanthropus (Peking man)	Hungary Germany China
Upper Pleistocene	100,000 yrs. to 50,000 yrs., 50,000 yrs. to 10,000 yrs.	Saldanha Neanderthal man Solo man Broken-hill (Rhodesian man) Cro-Magnon man	South Africa Asia, Europe, Africa Java Africa Asia, Europe
Holocene	10,000 yrs. to recent times	Differentiation of modern races of man: Australoids Caucasoid Mongoloids Negroid Bushmen Neolithic culture started 7000 yrs. ago – beginning of agriculture; domestication of animals; beginning of urban life.	

The cranial capacity of Australopithecine was around 600 to 700 cm^3 and that of Pithecanthropus or the Homo erectus found in China, Java etc. was around 900 to 1200 cm^3 as against that of Neanderthal man at around 1450 cm^3, which is practically near that of modern man. Pithecanthropus men were excellent hunters and probably knew about the uses of fire, particularly for cooking. They had very strong communal ties, probably engaged in warfare and practised cannibalism. In physical aspects, Cro-Magnon men were almost completely like modern man. They were culturally far more advanced than other primates of those periods and could perform all activities like modern men, except writing.

Neanderthal men made jewellery of distinctive design and style and wore them around their necks and arms.

Ever since the discovery of Australopithecus, it has been accepted that man, as we know him now, originated in Africa. He migrated to other countries during the period of glaciation – when much of the earth's water was in the form of great ice sheets, the level of the water in the oceans was lower than it has ever been and vast areas of the ocean floor were exposed and dry.

By the time the continental glaciers melted and the level of the seas rose, man moved to Europe, India, Southern Asia, Australia – and thus established his foothold on all known land masses on earth. At the same time, evolution was undergoing a decisive change. The development of the brain, intelligence and related adaptation progressed at a much faster pace than the evolution and adaptation of physical size and form. A study of man's advance in what could be classified as intelligent activity will lead us to understand how various civilisations evolved and took root, as we understand them now.

Chapter XXIV
CIVILISATION

With the evolution of man, his cultural history started. He seems to have started burying the dead as far back as 600,000 years ago. Their burial places also bear witness to the existence of some sort of rituals for the dead.

In some tribes, only chiefs were buried; in others, all members of the tribe were buried. Likewise, it seems that in some tribes only the very tall were selected as chiefs. Neanderthal and Cro-Magnon people practised some rituals at the time of burial, as evidenced by remains of food, necklaces, carved stones, etc. placed by the side of the body. It is presumed that they might have believed in some form of life after death. Cro-Magnon men used red ochreous earth to smear or sprinkle the dead body.

It was a common practice to keep the bodies in the flexed position, symbolic of a return to the foetal position, in burial. Could this practice be an indication that they had some kind of idea about rebirth and life after death?

Early human activities that could be related to the development of art and culture were graphic in character and found on walls inside caves or as rock incisions. Among the famous rock graffiti found are those in the Saharan region of Libya, Algeria and Tunisia. Communication seems to have been a prime objective of the graffiti indicating roads, street signs, etc. in Valcamonica, Italy. The use of arrows for marking roads was practised at least 10,000 years ago. Sculpture was another early art form. Sketches of females were used as symbols of fertility. With communication of ideas made possible, man became a liberalised animal. He was no more tied down by the dictates of the environment. This liberalisation kindled aesthetic desires and ideas were born. Satisfaction of the appetite was no longer sufficient. He started looking at his universe and wanted to see it closer and understand it in greater depth. Death, though frightening, was something to be understood and probed into. As seen earlier, almost 7,000 years ago man started many organised forms of activities such as agriculture, domestication of animals and collective living. His urge to understand the mysteries that surround death and birth; his eagerness to alleviate pain due to accidents or illness; his desire to imitate nature and communicate the pleasure and happiness derived by seeing and feeling nature; the feeling of importance and satisfaction of his ego on being able to exercise his authority on others around him, and the ecstasy when his sexual urge was satisfied, might have formed the basis for subsequent development and laid the foundations for the growth of civilisations.

We have records of well-developed civilisations, dating back to almost 4,000 years. Even though some documentary evidence of life before that period is available, it is sketchy and lacks continuity. Since river valleys were more congenial to collective living and agriculture, all the great civilisations originated in river valleys. The conditions in cities provided the right environment for philosophers, scientists, artists and writers. Architecture, literature, painting, music, philosophy and science flourished. The great civilisations that flourished can be grouped under five major geographical areas.

1. Middle East:	Mesopotamian
	Persian
	Greek
	Egyptian
	Arab
	Turkish
	Phoenician
	Hebrew
2. Europe:	Western
	Roman
	Minoan
	Byzantine
3. South Asia:	Indian
4. Far East:	Chinese
	Mongol
5. American:	Pre-Columbian
	Modern
	Colonial

Each of the major civilisations listed had its own distinct features. However, one aspect that needs special mention is that the two civilisations that survived from the very beginning to this day, namely the Chinese and Indian, are those based on philosophy, moral values and a way of ethical living in preference to military strength and power. They were able to withstand repeated attacks by many others of superior military strength, survive and re-emerge after every attack or after short periods of subjugation.

*Chart showing sequences of development and major events characterising various civilisations**

Yrs. ago	Geographical area	Known as	Major developments
4000	1	Mesopotamian	Sumerian city states; irrigation; extraction of copper; alloying of metals; writing; division of circle into 360 degrees.
		Egyptian	Irrigation; maths; solar year; writing; construction of pyramids.
	3	Indian	Dravidians in Indus Valley; irrigation; city communities; pottery; orderly system of government; trade with Mesopotamia, one of the two oldest living civilisations.
	4	Chinese	Agriculture; medicine; river conservancy; irrigation; rule of sage kings; one of the two oldest living civilisations.
3900	1	Egyptian	Writing on papyrus reed; period of Abraham; founding of Jewish nation.
	2	Minoan	Centred on Crete; prosperous trade; art; sculpture and painting; copper and bronze.

* The correlation between the period and the events is not necessarily absolutely correct; the intention is to present an approximate time frame of the events to give an overall picture.

Yrs. ago	Geographical area	Known as	Major developments
3500	1	Mesopotamian	Babylonian Empire; code of laws introduced.
	1	Hebrew	Early Hebrews; Anatolia; Efficient system of Govt. and law.
	3	Indian	Indus valley; Aryan invasion; Dravidian migration to south; intermarriage and fusion of races and culture.
	4	Chinese	Complicated form of character writing.
3200	1	Phoenician	Sailing skill; trade; invented 22-consonant alphabet.
	1	Egyptian	Architecture; temples at Thebes.
	1	Hebrew	Moses; Exodus from Egypt; Ten Commandments.
	1	Greek	Mycenaean civilisation; Indo-European invasion of Greece; absorb Minoan culture; siege of Troy; Dorain invasion.
	3	Indian	Early Hindus; epics; religious hymns; Rig-Veda.
	4	Chinese	Shang dynasty; lunar calendar; bronze vessels.
2900	1	Mesopotamian	Assyrian empire; military superiority; iron implements and weapons.

Yrs. ago	Geographical area	Known as	Major developments
	3	Indian	Growth of Hindu religion; concept of all-pervading universal soul (Brahman); belief in cycle of rebirths; caste system.
	4	Chinese	Canals; dams; use of bronze & ivory; jade works; silk manufacture; poetry & lyrics.
2700	1	Egyptian	Decline of Egypt.
	1	Hebrew	Growth of Judaism; concept of one God – the creator and the judge; concept of chosen people.
	1	Greek	Greek city-states; great epics – Homer's Iliad.
	5	Pre-Columbian	Olmec civilisation of Central America.
2500	1	Persian	Persian empire; Zoroastrian religion adopts God of light as supreme.
	1	Hebrew	Jerusalem destroyed.
	1	Greek	Golden age; birth of modern philosophy; science; Athenian democracy; growth of art and democracy; growth of art and literature; Pythagoras law.
	2	Roman	Kings expelled; republic established; constitutional experiments.

Yrs. ago	Geographical area	Known as	Major developments
	3	Indian	Buddhism – salvation through renunciation, withdrawal, meditation and strict morality – sublime peace – nirvana; Upanishads.
	4	Chinese	Golden age of philosophy; Lao Tse – founder of Taoism; belief in harmony and goodness of nature; Confucius; practical lessons of wisdom and righteous behaviour.
2200	1	Greek	Hellenistic empire; Alexander the Great; science; arts; philosophy; geometry.
	3	Indian	Ashoka; Mahabharata epic.
	4	Chinese	Great wall of China.
	5	Pre-Colombian	Zaptec Indians of South Mexico.
	2	Roman	Italy and Mediterranean under Roman rule.
1900	1	Hebrew	Jesus Christ; New Testament.
	4	Chinese	Buddhism reaches China; use of paper.
1600	2	Western	Spread of Christianity.
	3	Indian	Golden age – Hindu literature; art and science; Sanskrit as major developed language; astronomy; decimal system and use of zero; poetry & drama.

Yrs. ago	Geographical area	Known as	Major developments
	4	Chinese	Rapid spread of Buddhism.
	5	Pre-Columbian	Tiahuanaco Indians – Columbia.
1400	1	Arab	Muhammad; birth of Islam; Arab civilisation in Baghdad; Damascus & Cordova; militant Islam extends to other areas.
	2	Roman	Declines.
1200	2	Western	Growth of monastic orders; spread of Christendom; concept of holy Roman empire; church as guardian of culture; spread of spiritual outlook.
	2	Byzantine	Eastern Roman Empire; Greek Orthodox Church.
	3	Indian	Buddhism declines in India; Hinduism revives; Buddhism spreads to South East Asia; early Arab invasions of India.
	4	Chinese	Tang Dynasty; Golden age of learning; exchange of ideas with India and Arab world; academy of letters; block printing; magnetic compass; gunpowder invented.
	5	Pre-Columbian	Mayan civilisation in Honduras.
1000	1	Arab	Dominance in Central Asia.

Yrs. ago	Geographical area	Known as	Major developments
	2	Byzantine	Spreads into Russia & Balkans.
	5	Pre-Columbian	Toltec Indians.
800	3	Indian	Islam establishes roots; Hindu militancy.
	4	Chinese	Influence penetrates Japan.
	4	Mongol	Mongol empire; parts of China annexed; short-lived.
	5	Pre-Columbian	Height of Mayan civilisation; Aztec Indian civilisation; Dominates Mexico.
400	1	Arab	Influence in Europe decreases; continues to dominate India and east Asia.
	1	Turkish	Ottoman empire at its height; Luther & Calvin; Protestant movement; Era of explorations; Columbus in West Indies; Vasco da Gama in Africa & India.
	2	Byzantine	Decline.
	3	Indian	Mogul empire; synthesis of Indian & Islamic cultures; Jesuits land in India; Taj Mahal built; Western powers show their presence and fight among themselves on Indian soil.

Yrs. ago	Geographical area	Known as	Major developments
	4	Chinese	Ming Dynasty; drives out Mongols; renaissance of art & literature; philosophy; Ming code of law.
	5	Colonial	Incas rule from Columbia to Northern Chile; Temple of Sun at Cuzon; Efficient political system.
			Colonies established by European powers; Columbus in W Indies; Cabot in North America; Cortez in Mexico; Pizarro in Peru; Portuguese in Brazil; English in Virginia & Newfoundland; French in Canada; slave trade starts.
200	1	Arab	Islam declines as a power.
	1	Turkish	Revolts lead to modern Turkey.
	2	Western	Age of reason; industrialisation, scientific inventions; William Harvey – on blood circulation; Galileo's telescope; Darwin's law of evolution; Newton's law of gravity; Watt's steam engine; golden age of science; Philosophers – Descartes, Kant; industrial revolution; French revolution; Shakespeare; Voltaire; Goethe; Balzac.

Yrs. ago	Geographical area	Known as	Major developments
	3	Indian	British empire in India – India revolts.
	4	Chinese	Manchu dynasty; expands sphere of influence.
	5	Modern	Colonial period ends; USA and other nations independent; literature, science and economic activities at their height; George Washington; War of Independence; Abraham Lincoln; slavery abolished; civil war in USA.

Recent Developments

This period is generally known for movements for independence; liberalisation from colonial and imperial powers; establishment of communism in Russia, other European States and China; acceptance of the concepts of liberty, equality and fraternity; economic protectionism and later economic liberalisation; rise of military strength of various nations; World Wars I & II; large-scale destruction; atomic power and its proliferation; League of Nations and later United Nations; further military adventures restrained largely by collective voice of humanity; large-scale use of electrical power and introduction of atomic power; freedom to all erstwhile colonies; Indian independence and the concept of non-violent forms of agitation; Gandhi; rise of Arab and pan-Islamic nationalism; failure of communism in Russia and European countries; all-round human progress on the one side and abject poverty and sub-standard living conditions on the other side; population explosion; world population crosses six billion; environmental degradation due to man-made causes; rivalry between different ethnic communities in various parts of Africa; tendency for revival of fundamentalism and religious orthodoxy; tremendous achievements in science and information technology; space exploration; man lands on moon; able to take pictures of planets from comparatively closer ranges from space; man's journey to space and back; computers and electronic media; franchise; equal opportunities and empowerment of women; closer exploration of the universe and better facilities for quicker communication, entertainment and understanding of weather conditions; increased medical facilities and control of diseases; higher life expectancy.

However, the struggle for the evolution of a just and equitable social order within specified political and geographical boundaries on a global level continues.

Chapter XXV
ENVIRONMENT

As seen from the story of the evolution of mankind and the subsequent development and progress of civilisation, man is different from all other living species in that he is not content in living as per the dictates of the environment. The evolutionary process has bestowed on him a much higher level of intelligence to overcome the difficulties and constraints placed before him by nature – either by circumventing or by overpowering the natural forces. He has gone one step further to harness the natural forces and make use of them to work for him. While other species have adapted themselves to the natural conditions around them, man consciously tried to mould the environment to suit his requirement and taste. In other words, while all other living species are eco-friendly man is not; and consequently human activities create an imbalance in the ecology, which results in degradation of the environment.

Man's Activities

AGRICULTURE

As the population grew, the requirement for agricultural land increased and consequently large-scale deforestation became a practice. This led to the extinction of many wild animals and many more are on the verge of possible disappearance from the earth. This has caused a serious disturbance to the ecological balance. Further, the increased use of fertilisers, pesticides, weed killer, etc. leaves its residue in the soil and water, and even gets embedded in vegetables and fruits. This in turn becomes a health hazard for animals and human beings.

Housing

Secure and comfortable housing has become one of the essential requirements for modern man. The consequences of this are: increased requirement of land and use of agricultural land for housing, which in turn leads to further deforestation; increased requirement of construction materials like wood, cement, steel, bricks, paints etc., which in turn increases the pressure on forests, mining, industry, etc.

Urbanisation

Man has found that community life offers him many advantages and this has led to increased urbanisation. Further, many other social factors centred around him make city life more and more attractive and necessary. This results in problems like inadequate facilities for proper disposal of human excreta, as well as waste generated from food and other articles of day-to-day use. Improper disposal of these materials results in pollution of soil and water.

Comfort

Man's search for increased comforts has resulted in the use of many synthetic, natural materials and chemicals, such as synthetic fibres for clothing, soaps and detergents and insecticides, for hygiene, cosmetics, drugs, paints, dyes, plastic articles, paper, processed food, and so on. Man also started mining for various metals such as iron, aluminium, copper and so on. All these generated waste at the time of mining, processing or manufacture as leftovers and residues. Invariably, such wastes pollute and degrade the environment. Man has also invented a large number of machines for his comfort. Today, these machines are found in every house, office and public place. All machines need energy to run them and the most convenient form of energy in use is electric energy. Apart from energy, some of these machines, such as refrigerators, air-conditioners etc., use chemicals (known as chloro fluoro carbons – CFC) which when leaked out, escape into the upper reaches of the atmosphere, react with ozone and reduce the depth of the ozone layer. We have earlier learned that the presence of ozone in the upper atmosphere acts as a protective shield against ultraviolet rays from the sun. We are today threatened by a possible depletion of the ozone layer – one among the many man-made potential disasters.

INDUSTRIALISATION

Large-scale requirements of materials, implements and devices have led to industrialisation, aimed at mass production to meet demand. This has brought about a situation wherein large quantities of raw materials are required to be relocated and moved to the production centres. Further, this industrialisation has brought about the necessity of discharging large quantities of waste materials, generally known as effluents. On the one hand, the hunt for new raw materials puts more pressure on forests and land; on the other, large-scale dumping of waste materials makes the environment dirty and unhygienic, and on many occasions poisonous and hazardous to life around.

LOCOMOTION

Man's curiosity to explore and satisfy his social needs has taken him to many lands and people of many regions. Today, his activity covers the entire surface of the earth and, of course, beyond to space. Travel has become a part of his civilised life and, as seen earlier, he also needs to move large quantities of material from one place to another. This large-scale movement has been made possible through his inventions of trains, ships, automobiles and airplanes. All these need energy obtained by burning fuels such as coal, petroleum products (petrol, diesel, fuel oil), natural gas or biomass. As products of combustion (burning), these fuels give out apart from energy a lot of gases such as carbon dioxide, carbon monoxide, oxides of sulphur and unburned residual hydrocarbons, etc. Emissions of these gases along with soot, dust and noise pollute the atmosphere and have become a serious threat to health and the quality of life.

ENERGY

Modern civilisation and the comforts of life depend to a very great extent on the availability of energy at economically viable costs and in a conveniently usable form. Today, one of the main forms in which energy is consumed, apart from direct use as fuel, is electrical power. Electrical power is generated from various primary sources such as :

1. Fuels, like coal, petroleum, natural gas, and biomass. The use of all fuels results in one or other form of pollution.
2. Natural sources, like waterfalls, wind, solar, geothermal and ocean-based waves, tides etc. Of these, the use of waterfalls is well established and is being used on a large scale all over the world where such resources exist. Wind energy has also become increasingly popular though it is presently being tapped to a very limited extent. Presently, USA and India are the largest producers of wind energy. Solar and other forms are not yet found economically viable for large-scale generation. Geothermal generation is possible only in very selected places. Generation of electric power from these natural sources is the safest from the point of pollution.
3. Atomic energy is potentially a very big source and if sufficiently developed can solve the power needs for the

entire humanity in adequate measure. Even though the generation of atomic power is comparatively free from pollution, people are worried about the safety aspects and potential hazard from atomic radiation. Further, the installation of atomic power plants requires very large investments; and getting the required finance is a very big problem.

Thus, we see that the generation of power and use of energy is associated with pollution and problems of hazard.

WARS

In the history of man and the development of his civilisation, wars between people of different races, geographical and political regions, religious beliefs and economic groupings, have become endemic. The weapons used in these wars have been made more and more sophisticated, lethal and extremely dangerous – not only to the warring factions but to humanity as a whole. Man is today faced with the horrifying spectre of weapons of mass destruction such as nuclear bombs, chemical and biological warfare and so on.

Thus, it is not too difficult to understand that many activities of man, though aimed at his well being and comfort, also bring about problems, which are against his own long-term interests. Some of these man-made problems, if left unchecked, can result in the total annihilation of man. We have to be conscious of these dangers, and every individual should take on the responsibility to do whatever is possible within his reach to reduce problems of waste, pollution and degradation of the environment. It is a duty and a challenge before all of us.

Chapter XXVI

OUR OWN EARTH

The solar system, of which our earth is just one member, is just one of the numerous such solar systems, or galaxies as we know them. The universe, by our present understanding, consists of more than one billion galaxies, as detected by the Hale telescope in Mount Palomar. There are probably many more, which our most powerful equipments are not able to detect and our most precise analytical techniques are not able to spot. In fact, man has not yet succeeded in finding the outer limits of the universe; and as we explore further and further, we are given the impression of an ever-expanding system. Stars group together into clusters known as galaxies; the galaxy in which our sun is a member is called the Milky Way. This consists of 100 billion stars and looks like a flattened disk; actually, a flattened ellipsoidal globe of approximate diameter $10x10^7$ km. With our eyes we can see around 2000 to 2500 stars and with binoculars around 3500 stars. The various galaxies seem to have been placed along the spikes of a wheel and the whole system is in continuous rotation around a probable hub. Even if we travel at the speed of light (298,000 km per second) it will take around 100,000 light years for us to reach the farthest star in our Milky Way. Likewise, the frontiers of the universe as we know it lie around $6x10^{22}$ km away from us.

The most astounding factors about the universe are:

1. Its vastness – almost beyond our comprehension, a

limitless expanse.

2. The perpetual motion – each particle in the universe is in perpetual motion around its own hub at extremely high velocities; be it the electrons in an atom, the earth around the sun, the stars (including the sun) around the central point of the galaxy, and the galaxies around a probable hub in the universe.
3. The vast expanse or void containing practically nothing in between the agglomerates of material particles such as stars, planets etc. In fact, the volume of the interstellar space is so large in comparison to the total volume of material in the universe, that even if all material in the universe were reduced to an atomic state and thrown into the space, there would not be more than one or two atoms in a cubic meter of space! In other words, the space would still remain, for all practical purposes, empty.

Today, apart from powerful telescopes, astronomers can depend on spectro analysis of light rays coming from celestial objects. Analysis of light from a star can provide substantial information about its various physical characteristics, such as diameter, mass, surface temperature, chemical composition, rotation, velocity, magnetic field, etc. The radio telescope is yet another powerful tool in the hands of astronomers. In recent times, with the aid of radio telescopes man has detected the presence of agents such as carbon dioxide, ammonia, water, formaldehyde, etc. in solar systems other than ours in the Milky Way.

Despite the fact that today man has made his presence in space and has even landed on the moon, it is extremely doubtful whether he will ever be able to fully understand the distant and mysterious outer regions of the universe.

Is there life elsewhere in this universe? We still do not know. Perhaps, there are other planets in this universe like our earth on which life exists. These may be less or more developed forms of life than those on earth. In this context, it may be worthwhile to mention that the latest find of the astronomer C M Draconis is a star system about 57 light years away. Astronomers and astrophysicists are of the opinion that there may be a planet in this star system where life could exist. Irrespective of all these possibilities, these questions will remain matters of our curiosity, hypothecation and exploration for a very, very long time. For us, our earth will always remain precious. Our sustained development would depend on our ability to safeguard the earth from pollution and environment degradation. We have to keep this earth of ours free from all weapons of mass destruction. We have to protect all human beings from hunger, malnutrition and deadly diseases. To achieve these targets, we may have to check the growth in population and concentrate on the quality of life for every person on earth through planned families. This is only possible when we collectively resolve to end wars between nations and peoples; decide to love each other; and accept each other irrespective of race, region or belief as brothers and sisters. The well being of each one of us is dependent on the well being of every one around us – of every one on this earth.

In comparison to the almost limitless universe, are we able to comprehend, how small each one of us is? Just a small speck! Will this realisation make us humble enough to cast away our false pretensions of importance, egoism and hatred? Will it lead us to live a life of peace between man and man; a life dedicated to the betterment of all forms of life on earth; a life that loves this earth. This is not only our paramount duty but is also a challenge to every one of us. Can we learn to love all our brothers and sisters on this earth; can we learn to love every plant, every animal, bird and fish on this earth; can we learn to love every city and every village on this earth; can we learn to love every river, mountain and valley on this earth; can we love every highway, street and footpath on this earth? This is the challenge! Let each one of us resolve to accept the challenge, and to do everything possible within our reach, in spite of difficulties and constraints, to keep this earth as 'OUR OWN BEAUTIFUL EARTH' and to hand over the same to posterity, as beautiful as it is.